The Computer Owner's Tech Journal

AF824773

By Lynne Davis

First Edition

www.lynnedavis.com

The Computer Owner's Tech Journal

This Computer Owner's Journal Belongs To:

Name ____________________________________

Address __________________________________

City, State, ZIP___________________________

Phone ___________________________________

TABLE OF CONTENTS

SUBJECT	PAGE
Hardware	2
Printers and Scanners	12
CD Drive	13
Audio Interface	13
External Storage Drives	13
External Devices	14
Gaming Devices	15
Smartphones	17
Bluetooth Devices	19
WiFi Devices	21
Internet Accounts	23
Cloud Storage	25
Backup History	26
Software Applications	27
Operating System Keystrokes	43
Web-Based Applications	45
Email Accounts	46
Project Notes	47

COMPUTER HARDWARE

Manufacturer:	Desktop Laptop Tablet
Model:	Serial #:
Screen Size:	Processor:
Storage:	Memory:
Date Purchased: Cost:	Purchased From:
Accessories:	

OS Update History:		Maintenance History:	
Date	Desciption	Date	Description

Manufacturer:	Desktop Laptop Tablet
Model:	Serial #:
Screen Size:	Processor:
Storage:	Memory:
Date Purchased: Cost:	Purchased From:
Accessories:	

OS Update History		Maintenance History	
Date	Desciption	Date	Description

Manufacturer:	Desktop Laptop Tablet
Model:	Serial #:
Screen Size:	Processor:
Storage:	Memory:
Date Purchased: Cost:	Purchased From:
Accessories:	

OS Update History		Maintenance History	
Date	Desciption	Date	Description

Manufacturer:	Desktop Laptop Tablet
Model:	Serial #:
Screen Size:	Processor:
Storage:	Memory:
Date Purchased: Cost:	Purchased From:
Accessories:	

OS Update History		Maintenance History	
Date	Desciption	Date	Description

Manufacturer:	Desktop Laptop Tablet
Model:	Serial #:
Screen Size:	Processor:
Storage:	Memory:
Date Purchased: Cost:	Purchased From:
Accessories:	

OS Update History		Maintenance History	
Date	Desciption	Date	Description

Manufacturer:	Desktop Laptop Tablet
Model:	Serial #:
Screen Size:	Processor:
Storage:	Memory:
Date Purchased: Cost:	Purchased From:
Accessories:	

OS Update History		Maintenance History	
Date	Desciption	Date	Description

Manufacturer:	Desktop Laptop Tablet
Model:	Serial #:
Screen Size:	Processor:
Storage:	Memory:
Date Purchased: Cost:	Purchased From:
Accessories:	

OS Update History		Maintenance History	
Date	Desciption	Date	Description

Manufacturer:		Desktop Laptop Tablet	
Model:		Serial #:	
Screen Size:		Processor:	
Storage:		Memory:	
Date Purchased: Cost:		Purchased From:	
Accessories:			
OS Update History		Maintenance History	
Date	Desciption	Date	Description

Manufacturer:	Desktop Laptop Tablet
Model:	Serial #:
Screen Size:	Processor:
Storage:	Memory:
Date Purchased: Cost:	Purchased From:
Accessories:	

OS Update History		Maintenance History	
Date	Desciption	Date	Description

Manufacturer:	Desktop Laptop Tablet
Model:	Serial #:
Screen Size:	Processor:
Storage:	Memory:
Date Purchased: Cost:	Purchased From:
Accessories:	

OS Update History		Maintenance History	
Date	Desciption	Date	Description

PRINTER / SCANNER HARDWARE

Printer:		Serial#:
Model:		Software Version:
Driver:		Support:
Purchased:	Cost:	Notes:

Printer:		Serial#:
Model:		Software Version:
Driver:		Support:
Purchased:	Cost:	Notes:

Printer:		Serial#:
Model:		Software Version:
Driver:		Support:
Purchased:	Cost:	Notes:

Printer:		Serial#:
Model:		Software Version:
Driver:		Support:
Purchased:	Cost:	Notes:

Printer:		Serial#:
Model:		Software Version:
Driver:		Support:
Purchased:	Cost:	Notes:

CD DRIVE

Brand:	Model:
Serial#:	Software Version:
Speed:	Support:
Purchased: Cost:	Notes:

AUDIO INTERFACE

Brand:	Model:
Serial#:	Software Version:
Driver:	Support:
Purchased: Cost:	Notes:

EXTERNAL STORAGE DRIVES

Brand:	Serial#:
Model:	Software Version:
Capacity:	Support:
Purchased: Cost:	Notes:

Brand:	Serial#:
Model:	Software Version:
Capacity:	Support:
Purchased: Cost:	Notes:

EXTERNAL DEVICES

Brand:		Serial#:
Model:		Software Version:
Capacity:		Support:
Purchased:	Cost:	Notes:

Brand:		Serial#:
Model:		Software Version:
Capacity:		Support:
Purchased:	Cost:	Notes:

Brand:		Serial#:
Model:		Software Version:
Capacity:		Support:
Purchased:	Cost:	Notes:

Brand:		Serial#:
Model:		Software Version:
Capacity:		Support:
Purchased:	Cost:	Notes:

Brand:		Serial#:
Model:		Software Version:
Capacity:		Support:
Purchased:	Cost:	Notes:

GAMING DEVICES

Device:	Brand:
Model:	Software Version:
Driver:	Serial#:
Purchased: Cost:	Notes:
Accessories:	

Device:	Brand:
Model:	Software Version:
Driver:	Serial#:
Purchased: Cost:	Notes:
Accessories:	

Device:	Brand:
Model:	Software Version:
Driver:	Serial#:
Purchased: Cost:	Notes:
Accessories:	

Device:		Brand:
Model:		Software Version:
Driver:		Serial#:
Purchased:	Cost:	Notes:
Accessories:		

Device:		Brand:
Model:		Software Version:
Driver:		Serial#:
Purchased:	Cost:	Notes:
Accessories:		

Device:		Brand:
Model:		Software Version:
Driver:		Serial#:
Purchased:	Cost:	Notes:
Accessories:		

SMARTPHONES

Manufacturer:	Serial#:	
Model:	Software Version:	
Storage Capacity:	Service Provider:	
Purchased From:	Date Purchased:	Cost:
Accessories:		
Notes:		

Manufacturer:	Serial#:	
Model:	Software Version:	
Storage Capacity:	Service Provider:	
Purchased From:	Date Purchased:	Cost:
Accessories:		
Notes:		

Manufacturer:	Serial#:	
Model:	Software Version:	
Storage Capacity:	Service Provider:	
Purchased From:	Date Purchased:	Cost:
Accessories:		
Notes:		

Manufacturer:	Serial#:
Model:	Software Version:
Storage Capacity:	Service Provider:
Purchased From:	Date Purchased: Cost:
Accessories:	
Notes:	

Manufacturer:	Serial#:
Model:	Software Version:
Storage Capacity:	Service Provider:
Purchased From:	Date Purchased: Cost:
Accessories:	
Notes:	

Manufacturer:	Serial#:
Model:	Software Version:
Storage Capacity:	Service Provider:
Purchased From:	Date Purchased: Cost:
Accessories:	
Notes:	

BLUETOOTH DEVICES

Device:	Brand:
Model:	Software Version:
Driver:	Support:
Purchased: Cost:	Notes:

Device:	Brand:
Model:	Software Version:
Driver:	Support:
Purchased: Cost:	Notes:

Device:	Brand:
Model:	Software Version:
Driver:	Support:
Purchased: Cost:	Notes:

Device:	Brand:
Model:	Software Version:
Driver:	Support:
Purchased: Cost:	Notes:

Device:	Brand:
Model:	Software Version:
Driver:	Support:
Purchased: Cost:	Notes:

Device:	Brand:
Model:	Software Version:
Driver:	Support:
Purchased: Cost:	Notes:

Device:	Brand:
Model:	Software Version:
Driver:	Support:
Purchased: Cost:	Notes:

Device:	Brand:
Model:	Software Version:
Driver:	Support:
Purchased: Cost:	Notes:

WiFi NETWORK DEVICES

Internet Provider:	Account #
User Name:	Password:
Network Name:	Wifi Password:
Support:	Notes:

Internet Provider:	Account #
User Name:	Password:
Network Name:	Wifi Password:
Support:	Notes:

Internet Provider:	Account #
User Name:	Password:
Network Name:	Wifi Password:
Support:	Notes:

Internet Provider:	Account #
User Name:	Password:
Network Name:	Wifi Password:
Support:	Notes:

Internet Provider:	Account #
User Name:	Password:
Network Name:	Wifi Password:
Support:	Notes:

Internet Provider:	Account #
User Name:	Password:
Network Name:	Wifi Password:
Support:	Notes:

Internet Provider:	Account #
User Name:	Password:
Network Name:	Wifi Password:
Support:	Notes:

Internet Provider:	Account #
User Name:	Password:
Network Name:	Wifi Password:
Support:	Notes:

INTERNET ACCOUNTS

Vendor	Account #	Username	Password

Vendor	Account #	Username	Password

CLOUD STORAGE

Provider:	Account ID #
User Name:	Password:
Storage Capacity:	Account Fees:
Support:	Notes:

Provider:	Account ID #
User Name:	Password:
Storage Capacity:	Account Fees:
Support:	Notes:

Provider:	Account ID #
User Name:	Password:
Storage Capacity:	Account Fees:
Support:	Notes:

Provider:	Account ID #
User Name:	Password:
Storage Capacity:	Account Fees:
Support:	Notes:

BACKUP HISTORY

DATE	SOURCE DRIVE	DESTINATION	NOTES

SOFTWARE APPLICATIONS

Application:	Date Installed:
Brand:	Location:
Version:	Support:
Format:	Serial #:
Purchase Date: Cost:	License:
Notes:	

KEYSTROKES	
ACTION	KEYSTROKE

APPLICATION NOTES

WORKFLOW STEPS	
ACTION	INSTRUCTIONS

Application:	Date Installed:
Brand:	Location:
Version:	Support:
Format:	Serial #:
Purchase Date: Cost:	License:
Notes:	

KEYSTROKES	
ACTION	KEYSTROKE

WORKFLOW STEPS	
ACTION	INSTRUCTIONS

Application:		Date Installed:
Brand:		Location:
Version:		Support:
Format:		Serial #:
Purchase Date:	Cost:	License:
Notes:		

KEYSTROKES	
ACTION	KEYSTROKE

WORKFLOW STEPS

ACTION	INSTRUCTIONS

Application:		Date Installed:
Brand:		Location:
Version:		Support:
Format:		Serial #:
Purchase Date:	Cost:	License:
Notes:		

KEYSTROKES	
ACTION	KEYSTROKE

WORKFLOW STEPS	
ACTION	INSTRUCTIONS

Application:	Date Installed:
Brand:	Location:
Version:	Support:
Format:	Serial #:
Purchase Date: Cost:	License:
Notes:	

KEYSTROKES	
ACTION	KEYSTROKE

WORKFLOW STEPS	
ACTION	INSTRUCTIONS

Application:	Date Installed:
Brand:	Location:
Version:	Support:
Format:	Serial #:
Purchase Date: Cost:	License:
Notes:	

KEYSTROKES	
ACTION	KEYSTROKE

WORKFLOW STEPS

ACTION	INSTRUCTIONS

Application:	Date Installed:
Brand:	Location:
Version:	Support:
Format:	Serial #:
Purchase Date: Cost:	License:
Notes:	

KEYSTROKES	
ACTION	KEYSTROKE

WORKFLOW STEPS	
ACTION	INSTRUCTIONS

Application:	Date Installed:
Brand:	Location:
Version:	Support:
Format:	Serial #:
Purchase Date: Cost:	License:
Notes:	

KEYSTROKES	
ACTION	KEYSTROKE

WORKFLOW STEPS	
ACTION	INSTRUCTIONS

OPERATING SYSTEM KEYSTROKES

MAC	
ACTION	KEYSTROKE

OPERATING SYSTEM KEYSTROKES

WINDOWS	
ACTION	KEYSTROKE

WEB - BASED APPLICATIONS

Application:	Website:
User Name:	Password:
Brand:	Type:
Version:	Support:
Purchase Date: Cost:	License:
Notes:	

Application:	Website:
User Name:	Password:
Brand:	Type:
Version:	Support:
Purchase Date: Cost:	License:
Notes:	

Application:	Website:
User Name:	Password:
Brand:	Type:
Version:	Support:
Purchase Date: Cost:	License:
Notes:	

EMAIL ACCOUNTS

Provider:	Account ID #
Email Address:	
User Name:	Password:
Support:	Notes:

Provider:	Account ID #
Email Address:	
User Name:	Password:
Support:	Notes:

Provider:	Account ID #
Email Address:	
User Name:	Password:
Support:	Notes:

Provider:	Account ID #
Email Address:	
User Name:	Password:
Support:	Notes:

PROJECT NOTES

PROJECT:	
DATE:	APPLICATION:

PROJECT:

DATE: APPLICATION:

PROJECT:	
DATE:	APPLICATION:

PROJECT:	
DATE:	APPLICATION:

PROJECT:	
DATE:	APPLICATION:

PROJECT:

DATE: APPLICATION:

PROJECT:

DATE: APPLICATION:

PROJECT:	
DATE:	APPLICATION:

PROJECT:

DATE: APPLICATION:

PROJECT:

DATE: APPLICATION:

PROJECT:

DATE: APPLICATION:

PROJECT:

DATE: APPLICATION:

Made in the USA
Monee, IL
04 November 2026

67375590R00036